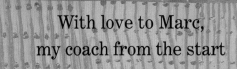

With love to Marc, my coach from the start

Special thanks to Maria Bettencourt
of the Massachusetts Department of Public Health

First Edition

Library of Congress Cataloging-in-Publication Data
Brown, Laurene Krasny.
  The vegetable show / Laurie Krasny Brown. — 1st ed.
    p.   cm.
  Summary: Presents information about nutrition and
vegetables, by means of pictures of singing, dancing vegetable
characters.
  ISBN 0-316-11363-8
  1. Vegetables— Juvenile literature.
2. Nutrition—Juvenile literature.   [1. Vegetables.   2. Nutrition.]
I. Title.
TX557.B76   1995
641.6'5 — dc20   93-25001

10 9 8 7 6 5 4 3 2 1

SC

Published simultaneously in Canada
by Little, Brown & Company (Canada) Limited

Printed in Hong Kong

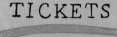

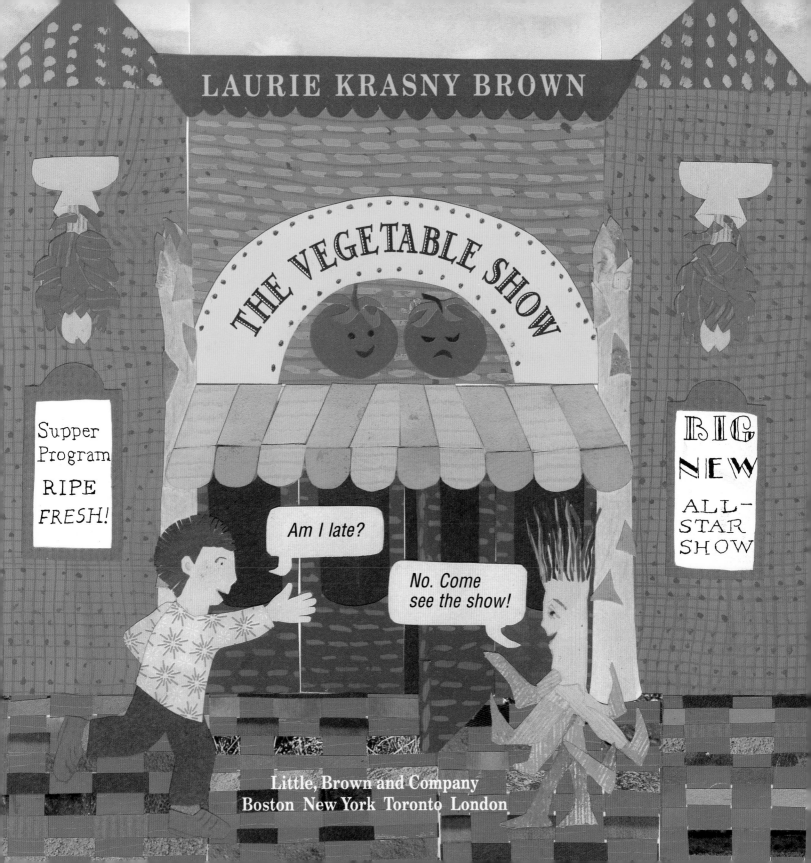

"Welcome to the Garden Street Theater!
I am your host," announced Mr. B. A. Dilly.
"Tonight we are proud to present . . .
Vegetable Vaudeville . . .
the Greatest, Greenest Show on Earth,
featuring a healthful helping of peak performers!
So sit back and enjoy the show."

Brassica
oleracea

Supper Program

The Garden Street Theater

SATURDAY EVENING

LITTLE, BROWN AND COMPANY'S SPECIAL PRODUCTION

THE VEGETABLE SHOW

by Laurie Krasny Brown

☛ Tonight's Performers in Order of Appearance:

À la Carte .... Host

B.A. DILLY .... Juggler

STRING BEANIE .... School Days

The Wee Peas *and* Ms. Shelly .... Magician

Bud the Spud ....
With his helper, Sweetie

LOTTA ROOT .... Singer
Accompanied by Mr. Gus

Eeny-Weeny Zucchini .... Weight Lifter

Last of the Red Hot Peppers .... Balancing Act

The TIP-TOP TOMATO TWINS ... Ballerina Acrobats

The VEGGETTES .... Chorus Line
Featuring Ruby Beet and the Dark Green Leafies

"For our First Act,
I have the pleasure of
introducing a most flavorful fellow,
the fresh and snappy . . .
Mr. String Beanie!
Mr. Beanie, just what are you doing?"

Beans are a tummy's friend.

SPRING SCHOOL
ALL LEGUMES WELCOME

"Next let's greet sweet
Ms. Shelly and the Wee Peas!
Protein-packed
and ready to pea-tend —
oops, I mean *pretend*!"

Early Peas

VINE COUNTY

Today's Riddles

Why shouldn't you tell secrets in a garden?
Because corn has ears and potatoes have eyes.

Which vegetable is the fastest?
The runner bean.

What's green and flies?
Super Pea!

Black-eyed Pea was here!

ELIZA

Me! Me!

Hmm ...

"For Act Three,
Bud the Spud
will mystify us
with his magic tricks!"
announced Mr. B. A. Dilly
from off stage.

See! I'm just a plain potato.
Nothing up my sleeve.
Now for the magic words—

Wait!
You forgot
your wand.

cook right

"Next I present the lovely
Miss Lotta Root,
with her twenty-four-karat voice,"
said Mr. B. A. Dilly.

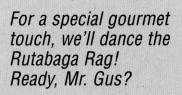

"Next on our stage is Eeny-Weeny,
the Amazing Zucchini!"
announced Mr. B. A. Dilly.
"Let's hear a rousing round of applause
for the world's strongest squash!"

Vegetables help make you strong!

Groan...grunt...ugh...up!

Yippee!

Wow!
He did it!

"For Act Six,
we will perk up your
appetites with these
sweet and spicy acrobats.
Let's welcome the
Last of the Red Hot Peppers,"
said our host.

Come on.
You can do it!

Uh-oh. This
could be
dangerous.

Sweet bell peppers, red, yellow, and green;

Perfect at meal times and in between.

Buon Appetito

"And now," said Mr. B. A. Dilly,
"for a *wow* finish,
here are . . . the vitamin-powered
Veggettes!"

taste
a
tomato

plant
a
pea

VEGGIES ARE NOT FOR SISSIES

With gusto

1. Veg - gies, ___ we're not for sis - sies, ___
Veg - gies, ___ we'll make you strong. And since we
come in ___ so man - y fla - vors, ___ We hope you'll
try us; you just can't go wrong!

2. Veggies, we're so delicious,
Veggies, we're such a treat.
And once you know that we're so nutritious,
We hope you'll try us; we just can't be beat!

pick
a
pepper

crunch
a
carrot

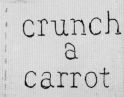

"*Lettuce* entertain you again soon!"
chorused Mr. B. A. Dilly and the Veggettes.

Bravo!
Bravo!

Hurray!

About the Performing Vegetables

Asparagus: (shoot) The asparagus stalk (called a spear) grows thick and juicy, and the small, pointed scales that grow on the stalk are called the leaves. The ancient Greeks and Romans used asparagus as medicine, to relieve toothaches and bee stings. Today we know that asparagus helps our bodies because it has vitamin A, vitamin C, and folic acid, a B vitamin. Asparagus is one of the first vegetables of spring.

Beet: (root) Beets are a "tame" variety of a plant known as wild chard, which still grows wild in southern Europe. Beets can be pickled, boiled and eaten with butter, or cooked in the Russian soup called borscht. The leaves are nice in salads. Certain sugar beets are so sweet that they are used to make sugar!

Carrot: (root) Carrots have vitamins A and C. Vitamin A helps our eyes adjust to dim light, keeps skin smooth and healthy, and promotes growth. Carrots are good raw, cooked, squeezed into juice, or baked into bread!

Cucumber: (fruit) This is one of the vegetables that have been grown by people for the longest time. Dark green cucumbers without soft spots are the tastiest. When the small, round kind of cucumbers are soaked in a spicy bath, they turn into pickles — like Mr. B. A. Dilly.

Green Bean: (seed pod) Native Americans were growing green snap beans long before Columbus arrived! Today there are many varieties, including runner, lima, string beans and a yellow type known as the wax bean or butter bean. A good source of vitamin C, vitamin A, and potassium, green beans are delicious nibbled raw or cooked.

Green Pea: (seed pod) Peas belong to one of the largest plant families, called legumes. Peas in the dried form were used in ancient times and have been found in Egyptian tombs. Legume seeds always grow inside pods made of two sections that fit together lengthwise. Snow peas, also called sugar peas or Chinese peas, with very tender shells, are eaten pod and all! Freshly picked green peas provide us with vitamin A, vitamin C, protein, and fiber.

Leafy Greens: (leaf) Leafy greens, such as lettuce, spinach, kale, and Swiss chard, are among the most healthful vegetables, rich in vitamins A and C and in the minerals potassium, calcium, and iron, and high in fiber. The darker the leaf's color, the more nutrients it has. Leafy greens are grown all over the world, but there is an American city that calls itself the world's spinach capital: Crystal City, Texas.

Pepper: (fruit) Sweet bell peppers taste mild and can be eaten raw or cooked. Hot, or chile, peppers are used as seasoning to make food taste spicy. All peppers will turn from green to their ripe color (yellow, orange, purple, or red) if left long enough on the plant. Peppers are excellent sources of vitamin A and vitamin C.

Potato: (tuber) The potato is the world's most widely consumed vegetable. Potatoes grow underground, growing larger and larger until they are full size. They also grow eyes, or buds, which can produce new potato plants. A simple baked potato with its skin is a good source of many nutrients, including vitamin C, some B vitamins, minerals such as iron, and fiber.

The sweet potato is not related to the ordinary potato. It is a tuber on a special kind of morning-glory vine. It is said to be the most healthful vegetable of all, rich in vitamin A as well as vitamin C and fiber.

Tomato: (fruit) Tomatoes are a good source of vitamin C. Vitamin C helps our bodies heal cuts and broken bones, keep gums healthy, and resist infections. You can have them as juice or cooked into sauce for spaghetti. Tomato lovers say that the most delicious tomato is the one just picked off the stem — sweet and ripe!

Zucchini: (fruit) A type of summer squash, zucchini is another vine fruit that we use as a vegetable. *Zucchini* is an Italian word that means "little squashes." Best from late spring to early fall, choose zucchinis that are small, firm, and smooth. Try them uncooked, cut into rounds or long, skinny spears good for dips.

Did You Know?

● Vegetables, like fruits, are the edible parts of certain plants. They are grown to be eaten.
● Cucumbers, peppers, squash, and tomatoes are all fruits that have been reclassified as vegetables because people eat them in the main part of a meal.
● The United States Department of Agriculture recommends that we all eat three to five servings of vegetables daily to stay healthy and grow properly. This is because of the important vitamins, minerals, and fiber vegetables supply.
● Although most common vegetables are available year round, they tend to have the best food value, flavor, texture, and price when in season.
● Cooking most vegetables as briefly as possible preserves their taste and freshness. Long cooking also destroys some vitamins. Try eating vegetables raw — they're crunchier that way!